Royal Canadian

ROYAL
CANADIAN

1906-07

Illustrated Catalogue

Royal Canadian

Rubber Footwear

Supreme Style

The Canadian Rubber Company
of Montreal Limited

"LEADERS AND ORIGINATORS"

SEASON 1906-7

ROYAL CANADIAN
THE CANADIAN RUBBER CO OF MONTREAL LIMITED

PHOTO ALLEMÃ   MANÁOS 1904-05

CARRYING RUBBER "MILK" TO SMOKING PLACES,
MANÁOS, SOUTH AMERICA

Royal Canadian

Our regular makes of Rubber Footwear are noted for many exclusive features.

"Canadians" are the dominant factor in the Rubber Footwear business of the Dominion.

Resourcefulness in designing, honesty in manufacture, aggressiveness in selling, are a triumvirate of reasons for our position at the top. And yet we're not satisfied.

Continual vigilance for further possible improvement is one of our strong tenets for 1906.

This Catalogue illustrates one of our latest developments in Rubber Footwear—the "Royal Canadian."

Here's the kernel of our proposition: We couldn't improve the wearing quality of our regular brands of Rubbers, so we decided to manufacture a special brand, carrying highest grade linings, finest finish—and noted above everything else for Style.

The price will be a little higher than any other Rubber on the market. But the price is a small factor in a comparative sense with the value comprised in the "Royal Canadian."

There is one quality in a Rubber that is hard to describe. It is Style.

Some Rubbers have it, but most Rubbers lack it. "Royal Canadians" are first amongst stylish Rubbers.

Every pair of "Royal Canadians" are packed in high grade cartons.

We illustrate in the following pages some leaders in the "Royal Canadian" grade.

The Canadian Rubber Company

Season 1906-7.   of Montreal Limited.

"LEADERS AND ORIGINATORS."

D. LORNE McGIBBON,
General Manager.

CLOTH GOODS

English Nettings and Jersey are noted for high finish and exclusive wearing features.

The Jersey used in all our Cloth goods is the product of the most noted looms of Nottingham, England. It is "jet black," and will not change colour through all kinds of wear and weather.

"Royal Canadian" Brand of Cloth goods are supremely handsome and shapely on the feet. The linings and all details embody superb style, and the lasts used are models of elegance and exclusive design.

ALL GOOD SHOE STORES SHOULD CARRY A FULL LINE OF "ROYAL CANADIANS" FOR THEIR MOST DISCRIMINATING TRADE. : : :

MIKADO

Extra Light Jersey Excluder.   3 Buckles.

Packed in Cartons.   12 Pairs to Case.

STYLE OF TOES

Men's........ .....Ideal, Empire, Crown, London.

LIST PRICE

Men's $3.25

EMPEROR

Extra Light Jersey Artic.

Packed in Cartons.   12 Pairs to Case.

STYLES OF TOES

Men's...... . .....Ideal, Empire, Crown, London, Tuxedo.

LIST PRICE

Men's $2.00

EMPRESS

Extra Light Jersey Gaiter.   Two Straps and Buckle.

Packed in Cartons.   12 Pairs to Case.

STYLES OF TOES

Women's.......Ideal, Empire, Crown, London, Opera.

LIST PRICE

Women's $2.50

QUEEN

Extra Light Jersey Buttoned Gaiter.

Packed in Cartons.   12 Pairs to Case.

STYLES OF TOES

Women's .....Ideal, Empire, Crown, London, Opera.

LIST PRICE

Women's $2.40

TAPPING THE RUBBER TREE, AND PLACING THE CUPS,
MANÁOS, SOUTH AMERICA.

CZAR

Extra Light.   High Front.   Light Jersey Upper.

Packed in Cartons.   25 Pairs to Case.

STYLES OF TOES

Men's ....Ideal, Empire, Crown, London, Tuxedo.

LIST PRICE

Men's $1.50

CZARINA

Extra Light.   High Front.   Fine Jersey Upper.

Packed in Cartons.   25 Pairs to Case.

STYLES OF TOES

Women's.. ....Ideal, Crown, Empire, London, Opera, Tuxedo.

LIST PRICE

Women's. $1.25

"The Raw Material of a Great Industry."

CUTTING, ASSORTING AND PACKING OF PARA RUBBER
AT MANÁOS, SOUTH AMERICA.

Pure Gum Specialties

The fine lines of a Rubber give the handsome appearance expressed by the word "Style."

"Royal Canadians" are supreme in the beauty of their appearance.

The linings throughout are Royal Purple, and every detail is planned for stylish effect.

"Royal Canadian" Pure Gum Specialties are exclusively the best in the world.

ASK TO SEE SAMPLES AND PLACE YOUR ORDER EARLY

Your Jobber carries a full line.

PRINCESS

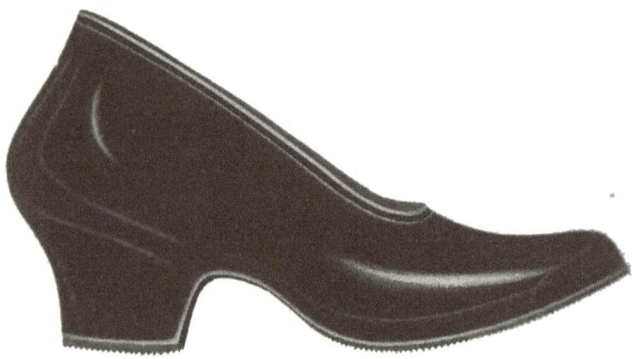

An Extra Light Pure Gum Croquet.   Will fit the full Louis XV French Heel.

Packed in Cartons.   25 Pairs to Case.

LIST PRICE

Women's ....$.85

VICTORIA

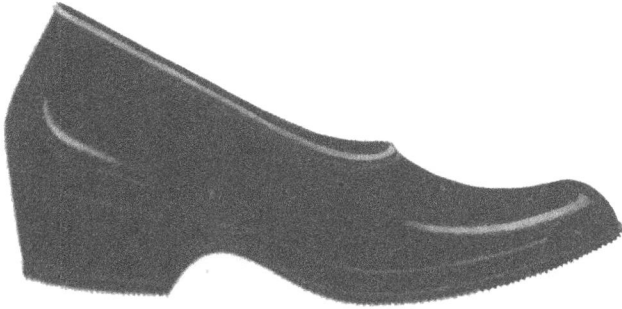

Extra Light Pure Gum Croquet.

Packed in Cartons.   25 Pairs to Case.

STYLES OF TOES

Women's.. ....Ideal, Empire, Crown, London, Opera, Tuxedo.

LIST PRICE

Women's.... ....$.85

ROYAL
THE CANADIAN RUBBER CO
OF MONTREAL LIMITED
CANADIAN

DUKE

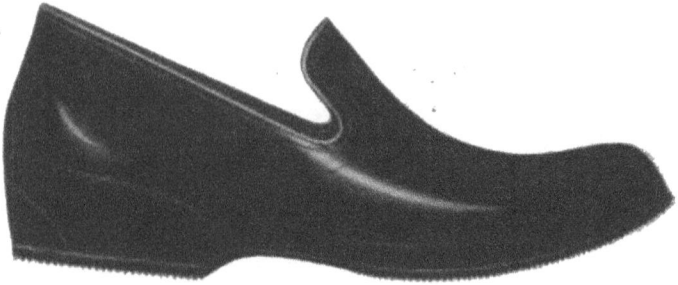

Extra Light.   Pure Gum.   High Front.

Packed in Cartons.   25 Pairs to Case.

STYLES OF TOES

Men's...........Ideal, Crown, Empire, London, Tuxedo.

LIST PRICE

Men's......... .... $1.25

DUCHESS

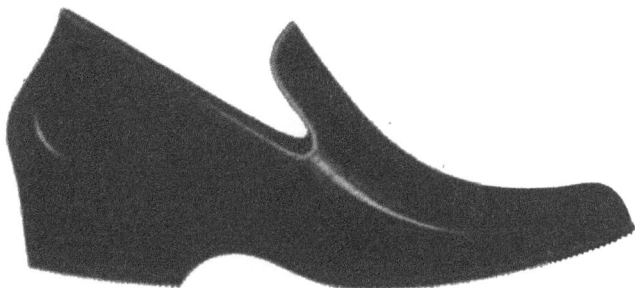

Extra Light.   Pure Gum.   High Front.

Packed in Cartons.   25 Pairs to Case

STYLES OF TOES

Women's .... Ideal, Empire, Crown, London, Tuxedo.

LIST PRICE

Women's $1.00

BARON

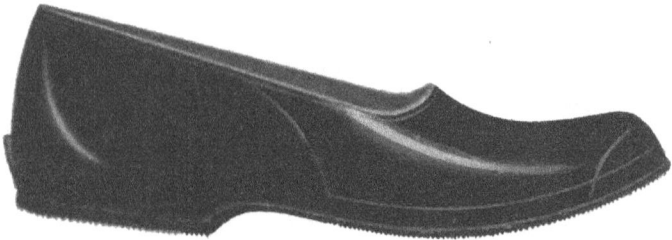

Extra Light Pure Gum, Self Acting.

Packed in Cartons.

25 Pairs to Case.

STYLES OF TOES

Men's..............Ideal, Empire, Crown, London, Tuxedo.

LIST PRICE

Men's...................... ....$1.15

ROYAL
THE CANADIAN RUBBER CO
OF MONTREAL LIMITED
CANADIAN

A Typical Scene in One of the Great Rubber
Producing Centres of the South

CUTTING, ASSORTING AND PACKING OF PARA RUBBER
AT MANAOS, SOUTH AMERICA

ROYAL
THE CANADIAN RUBBER CO.
OF MONTREAL LIMITED
CANADIAN

THE HOME OF "CANADIAN" RUBBERS

Canadian Rubber Co.
OF MONTREAL
NEW OFFICES & WAREHOUSES

Main Warehouse and Executive Offices: MONTREAL, QUE.

VIEW OF FACTORIES, MONTREAL, QUE.
Present Floor Area--21 Acres.

Royal Canadian

Supreme Style